HON. WILLIAM H. SEWARD'S

ARGUMENT

ON AN IMPROVEMENT IN THE

STEAM ENGINE.

THE STEAM ENGINE.

HON. WILLIAM H. SEWARD'S

ARGUMENT,

IN THE

CIRCUIT COURT OF THE U. STATES,

AT

COOPERSTOWN, NEW YORK,

ON THE THIRD DAY OF AUGUST, 1853.

Mr. Justice Nelson presiding, in the Cases in Equity, William B. Sickles and others vs. *George H. Corliss and John Barstow, and the same Complainants* vs. *Young and Cutter.*

PROVIDENCE:
KNOWLES, ANTHONY & CO., PRINTERS.
1853.

The Complainants in these cases had moved for preliminary injunctions against the Defendants, to restrain them from making, vending and using steam engines which contained an improvement alleged to be an infringement of a cut off, claimed to have been invented by Frederick E. Sickles, of New York, and patented by him in May, 1842. The Defendants denied the infringement, and produced patents granted to George H. Corliss in March, 1849, and in July, 1851, for the improvements in the machinery built and used by them, and they claimed that George H. Corliss was the first original inventor of these improvements, and that they were substantially different from the alleged invention of Sickles, and greatly superior in value to his supposed improvement, and to all other modes of regulating the velocity of the steam engine, and of working steam expansively. The Defendants also denied that Sickles was the inventor of the improvement patented by him.

The causes were heard upon the bill, answers and affidavits filed by the several parties, and was argued on the part of the Complainants by Charles M. Keller and Edward N. Dickerson, of New York; and on the part of the Denfendants by Hon. W. H. Seward, of Auburn, and Samuel Ames, of Providence, (with whom were associated Samuel Blatchford of Auburn and Thomas A. Jencks, of Providence,) for the Defendants.

The steam engines in question were built by Messrs. Corliss & Nightingale, of Providence, and the merits of the inventions which give peculiar value to these machines are fully set forth, and discussed in the following argument delivered by Mr. Seward, on behalf of the Defendants in the above named cases.

ARGUMENT.

May it Please the Court: Recent publishers of Tredgold's Treatise, the most recondite that has ever been written on the steam engine, invoked royal patronage by this significant dedication:

TO HER MOST GRACIOUS MAJESTY,

VICTORIA,

QUEEN OF GREAT BRITAIN AND IRELAND,

THESE VOLUMES,

ON THE PRINCIPLES AND CONSTRUCTION OF THE

STEAM ENGINE,

INVENTED AND PERFECTED BY THE TALENT AND ENTERPRISE

OF BRITISH SUBJECTS,

And Essential to the Progress of Civilization in every Nation,

ARE INSCRIBED.

It annoys and mortifies an artisan, prosperously engaged in manufacturing steam-engines, and erecting them at Boston, Providence, Hartford, Trenton, Harrisburg and other capitals, and in Lawrence, Lowell, Lancaster, New York, Philadelphia, and other seats of manufacturing industry, to be brought before this Court upon charges of trespass and piracy. Nevertheless, the occasion may prove to be one of advantage, if it shall appear in the end that not only is he not at all a trespasser, nor a mere artisan, but also that he is a philosophical and beneficent inventor, who has, like Fulton, by real and eminently useful additions to the perfection of the steam-engine, so confidently claimed for British subjects, won new distinction for American genius, and given a fresh impulse to the progress of civilization throughout the world.

Your Honor knows well the history of the steam-engine; its embryo in the rude device of the Marquis of Worcester, in 1663, for raising water to a higher level, which drew from that pious philosopher an ejaculation of "Thanks to God, next to those which are due for creation and redemption, for having vouchsafed an insight into so great a secret of nature, beneficial to all mankind, as this water commanding engine;" the invention by Savery of the atmospheric steam-engine, and its application to a hydraulic purpose; the accidental, though happy discovery, by Newcomen, of the process of condensation by the cold-water jet; and the rude state in which Watt found, and the improved condition in which he left, this great mechanical agent of Human Progress, when he had defined the due proportions of the boiler and the cylinder, had disclosed the laws of latent heat, and of the pressure of steam, had invented sliding and circular valves, and perfected steam-packing, had replaced the atmospheric with the double acting steam-engine, securing in it a perfect vacuum by a process of separate condensation, and new economy by working the expansive force of the steam, and had subjected the engine to imperfect yet useful regulation by the governor. And your Honor well understands, that it was when the steam-engine had attained this degree of perfection, that, as a stationary agent, it went into general use in the manufactories of Europe, while our own Fulton seized and converted it into a marine power, and sent it abroad to revolutionize the commercial, social and political systems of all nations. Our researches, on this occasion, will call us back to that period.

The cost of fuel is the chief impediment to the universal use of the steam-engine. Every improvement of it reduces this cost. Watt's inventions effected a saving of three-fourths of the fuel before fed to the engine, and he realized an ample reward by reserving to himself the value of one-third of that saving, although there was then not one steam-engine, where a hundred are now employed in the United Kingdom. The recently renewed, and I trust hopeful efforts of Ericsson to substitute heated air for steam, illustrate the importance of economy of fuel in the use of the steam-engine. Although the expense of the fuel is borne at first by the manufacturer, it ultimately falls on either the producer of the material or the consumer of the fabric; and so, economy in this respect is a great interest of society.

We bring here into Court the steam-engine as it has been perfected by Mr. George H. Corliss, and claim for it that it is far more effective than it was as left by Watt, while it saves more than one-half of the quantity of fuel that it then demanded. Mr. Corliss's improvements have thus far, however, been applied only to stationary engines; and, of these, only to such as use valves having a sliding or circular motion.

His improvements are two-fold:

I. A peculiar device for moving each steam and each exhaust valve, with a distinct and independent motion, by means of a crank-wrist. There being thus a series of crank wrists, all of which are attached to a common disc or plate, which is secured to a rock-shaft connected with the main eccentric; so that all of the valves are moved, though by distinct motions at the right times, and in the most advantageous manner. Each wrist operates, through a distinct lever, upon its proper valve; and all of the wrists are so arranged on the common wrist-plate, with reference to their levers, that they act like cranks, each of which vibrates near its *dead point*, or point of slowest throw, and therefore imparts but little movement to the valve it actuates when that valve is closed, while each moves with its fastest throw, and therefore imparts the greatest movement to its valve, when that valve is being opened.

This device is substituted for the common process, in which a steam valve and an exhaust valve are rigidly connected together, so that when one is moved the other is forced to move equally with it, and the closed valve is of course moved with the opening one, and consequently the whole amount of force consumed in thus unnecessarily moving the former while closed is expended to no good purpose, and tends to increase the wear of the engine.

This new device secures two advantages: 1st, it saves much of the power which was injuriously expended in moving the closed valve; 2d, it prevents wire-drawing or waste of the expansive force of the steam, because the valves are moved with increased speed while opening and closing their ports.

II. The second improvement in Mr. Corliss's engine consists in a method of perfect automatic regulation of the steam in its passage into the cylinder, so that by means of a cut-off at the steam valves, only, the entire expansive force of the steam is saved and applied. This is effected by combining the governor, all its sensibility being completely preserved through the agency of stops or cams, with the catches that liberate the steam valves for the purpose of cutting off the flow of steam into the cylinder.

Mr. Corliss's engine, then, is the stationary steam-engine with sliding or circular valves and with a new gear, adapted to perfect automatic regulation and to the saving of fuel by applying the entire expansive force of the steam. If it is useful at all, the public have a deep interest in it, because stationary engines are coming now into more general use than heretofore in every department of the arts—no less than two thousand being in operation on Manhattan Island alone.

The usefulness of Mr. Corliss's engine is implied by the conduct of the Complainants in regard to it. They have already brought two actions against those who were making or using it in the District of Rhode Island, one in the District of New Jersey, and three, of which that in which we are now engaged is one, in the Southern District of New York.

The utitility of the engine can be established, moreover, by scientific authority :

Tredgold, Art. 524—"The action of a steam-engine is variable; consequently, when an equable motion is necessary, its action must be equalized."

Art. 535—"An equable motion is desirable in almost every kind of machine, it being strained much more by an irregular, desultory one than when the motion is equable. * * * * While the strength of the machine must be adapted to the greatest strains that occur, the quantity of work done is equivalent to the mean action only, and more is not performed by a desultory motion than by one at a mean rate and uniform."

Art. 543—"An engine is frequently applied where the work to be done is not constantly the same; and when the machinery of a part of it is suddenly stopped, or suddenly sent on, if the moving power were to remain the same, an alteration of the velocity must take place—it must move faster or slower. This change of velocity would, in most cases, be very hurtful to the work, and cause considerable loss. Besides, there is always a velocity at which a machine must act with greater advantage than at any other. Therefore, the change of velocity arising from the above cause is in all cases a disadvantage, and in all delicate operations exceedingly injurious."

Art. 545—Concerning the regulation by the action of the throttle valve, (which is retained in Sickles's stationary engine and is discarded in the Corliss Engine) "An axis [throttle] valve of this kind has much advantage over a valve of any other form for a circular pipe, because it contracts the aperture without being difficult to move, or presenting more than the necessary obstruction. But it is by no means an economical mode of varying the power of the steam-engine."

The utility of Mr. Corliss's improvements admits of easy and simple demonstration. The regulation in his engine, differing from all others, is *purely automatic* and *practically perfect*.

The object of all mechanism is to substitute perfectly accurate mechanical action for the functions performed by animal power, and by human labor, skill and care. The regulation of the power of the steam-engine is always to be made with reference to two things: 1st, the change of *heat* in the furnace; 2d, the *load*, which varies with the different degrees of force exercised under circumstances ever-changing by the machinery propelled by the engine. It is manifest that the engineer can only conjecture the extent of the changes in these conditions affecting the motion of the engine, and therefore can adjust the regulation only with proximate accuracy. But in Corliss's engine, the adjustment is always made by the governor. The governor is the very *sensorium* of the engine, and must, if properly organized and effectively combined, perceive the changes of condition in the very instant of their occurrence, and adapt the regulation immediately and with full effect. So that the engine, instead of running at varying velocities, as it does when the adjustment is made by hand, runs practically at the same velocity through all changes in the conditions by which its motion is affected.

Observe the nicety of this automatic adjustment of the point of cut off in Corliss's engine. The adjustment takes place twice for every revolution of the main crank. In the the largest engines, those which make thirty revolutions in a minute, the adjustment of course occurs sixty times in a minute; and in small engines, making sixty revolutions, it occurs one hundred and twenty times in a minute. It needs no demonstration to show that delicacy of adjustment, like this, could not be attained by the human hand, nor has it ever before been attained in any mode of regulating the engine. Mr. Corliss, then, has the merit of having *perfected* the mechanism for regulating the power of the steam-engine.

Observe, also, the great excellence of the process of regution in Corliss's engine. Watt invented a method of regulation by combining the governor with the throttle valve and that combination has prevailed in nearly all the engines since and is found in Sickles's engine. Corliss, on the contrary, uses no throttle valve and effects the regulation altogether by combining the governor through the agency of stops or cams with the steam valves.

When the governor is combined with the throttle-valve, it is in fact converted into a connecting medium, and made not merely to indicate the point of adjustment, but also to perform the labor of moving the valve; and it has to perform this duty after reaching through steam-tight packing. This labor exhausts its power while it makes its proper function—that of a regulator merely subsidiary to its office of a motive lever.

Now, it was just for the reason that the elastic force of the steam was greatly diminished by the combination of the governor with the throttle valve, that some new device, with appropriate mechanism, was necessarily introduced, to regulate by means of a cut-off at the steam valves, and so graduate the supply of steam according to the wants of the engine, and then it was found necessary to return to the throttle valve to restore the sensibility of the governor. Among such devices were those in Maudslay's engine, described in Tredgold, at page 262, and certain French engines which have been mentioned here. None of these devices, however, attained a satisfactory automatic regulation. They could not attain it for the reason that either they wasted the expansive force of the steam in conveying it through the throttle valve or else, abandoning the throttle valve, and regulating altogether by a cut-off at the steam valves, they converted the governor into a lever to move the steam-valves and so far exhausted its sensibility that an adjustment by hand was necessary as it had been before.

These inconveniences resulted from a conflict of antagonistic elements in the engine as thus organized. The governor must necessarily be throughly senstiive; and, whether the regulation was to be effected at the throttle valve, or at the steam valves, if the two duties, to wit, of regulation and of actuation, were assigned to one agent, either the sensitiveness of that agent must be reduced or its power must be impaired; and so, the result was at best a compromise.

So, again, a cut-off cannot accurately produce the desired effect of regulation, unless it is adjusted by the governor. When a governor works a throttle valve or the steam valves, it either imparts or it resists a force, and this so far exhausts its sensibility that it cannot accurately fix the point of adjustment. Or, if its sensibility is preserved and the valves are worked independently of the governor, the cut-off is not regulated without great sacrifice of steam, according to the demands of the engine. And thus under the one arrangement of regulating at the steam valves the superintendence of the engineer is necessarily invoked, while in the other—that of regulating at the throttle valve—there is a waste of the expansive force of the steam.

In Corliss's engine, on the contrary, there are no antagonisms, and so there are no compromises. The governor is simply a regulator, the valves being moved by a force distinct from it, yet subjected to its regulation; thus, the regulation is not only adjusted according to every momentary change in the demands of the engine but is also effected with absolute precision by agencies purely automatic.

All this is seen in the model here. The governor no-where performs any labor, and, on the contrary, only indicates the change required to the levers which move the valves. This does not task its powers. It puts forth only the force necessary to move either a small cam or a stop. This movement is attended with the least possible friction and the cam or stop presents absolutely no resistence to the governor except at the very instant when it is in actual contact with the lever constituting its fulcrum. This momentary resistance by the bearing of the lever on the stop as a fulcrum occupies in an engine which makes sixty revolutions in a minute, only the fractional part of half a second, and this period compared with that during which the governor is left free to move the stop or cam is practically infinitesimal.

Again, the stop or cam which constitutes the fulcrum of the lever that moves the valves in the process of detaching them, is not rigidly attached to the mechanism, but is separate and free, and yields to the slightest touch of the finger of the governor.

Once more, the traverse of this adjustable stop or cam is inconsiderable; ordinarily less than half an inch; and such is the sensibility of the governor in this combination that this traverse may be reduced by leverage within infinitesimal limits. It is manifest, therefore, that the adjustment is absolutely perfect. The movements of even the strongest combinations may be interrupted and broken by an almost inconsiderable force, and they may be promptly restored again by delicate mechanism when the interrupting agent is withdrawn. The improvements of regulation in the steam engine which I have described have been made by carrying out this principle with great attention and constancy. These improvements are exactly an equivalent in the steam engine to the watch made by substituting the lever in the place of the improvement in the old crown wheel. In the old mechanism, the governor performed the duties not only of the pilot but of the man at the wheel also; while in Corliss's engine the governor presides and directs only, and the changes are made by the lever as promptly as their necessity is indicated and extent defined.

We pass now to the parts of Corliss's engine which are concerned in working by the whole expansive force of the steam in the cylinder. The use of the throttle valve in the old engines—including that of Sickles's—was always attended by a wire-drawing of the steam. This wire-drawing is, in other words, a reduction of the expansive force of the steam; and so, every form of cut off before used in engines having sliding or circular valves, for the purpose of working by that expansive force, was an imperfect one. But Corliss's arrangement dispenses with the throttle valve altogether and causes the steam valves to be opened suddenly; and thus the steam while admitted into the cylinder, has practically the same tension as that which it had when generated, and the whole expansive force of the steam is saved and used.

In puppet valve engines the valves must be started from their places of rest at the moment of closing or opening their ports. In Corliss's engine, on the contrary, the sliding or circular valves have a rapid motion at that point, resulting from the fact that their motion begins before the ports are covered or uncovered, and so he gains some advantage in this respect. He gains, however, a crowning advantage by his automatic method of varying the *periods* of closing the steam port—that is to say, the periods of completing the cut off. By means of these advantages Corliss's cut off becomes a perfect one. On comparing it with the engines using the throttle valve for regulation we see this difference. In the engines regulated by the throttle valve, the steam valves close always, at a fixed point, whatever may be the pressure of the steam or the load on the engine. If, then, we suppose the pressure to remain the same, and the load to be reduced one half, as the same volume of steam must enter the cylinder, the pressure of the steam must be reduced by a strangling process at the throttle valve to prevent excessive motion. Again, if the pressure be reduced while the load on the engine remains the same, then we fail to obtain the necessary velocity.

Corliss meets these emergencies thus: In the first case, by closing the ports the instant one-half the volume of steam has entered the cylinder; and, in the other, by keeping the port open until double the quantity of steam has entered the cylinder.

Again, the new arrangement prevents a wire-drawing of steam at the closing of the valve, by means of the suddenness of the motion; a result which cannot be attained in an engine having puppet valves, because the descent of the valves by gravity must, in such engines, be checked at some point to prevent their slamming on their seats. This checking of the descent of the valves has been effected by means of a piston, a plunger and a dash pot, of which we shall have occasion to speak more fully hereafter. But in Corliss's engine no such retardation takes place, or is necessary. The valves are suddenly brought over their ports, and so all wire-drawing at the closing of the valves is prevented, and the whole expansive force of the steam is secured. Here, also, a compromise between antagonisms found in all other engines is avoided. It is important in all engines to save the whole tension of the steam; but it is equally important to prevent the concussion of the valves against their seats. Both objects are secnred in Corliss's engine; and this, also, is a new and eminently useful result. To sum up the matter, the regulation of the engine is made perfect by the peculiar way of combining the governor with the cut off, and the cut off is made perfect by the automatic adjustability secured by that connection.

Authorities on the steam engine show the importance of this new result:

Tredgold, page 337—"The apparatus for opening and closing the passages is of more importance to the perfection of the steam-engine than any other part of its mechanism. In the present state of the engine the action is either very complicated or imperfect."

Page 204, Art. 435—"When valves, cocks or sliders are to be moved to admit steam to a steam-engine, the motion should be as quick as circumstances will admit, so that the passages may be wholly opened or wholly closed at the proper time with the least delay, for it may be easily shown that a considerable loss of effect arises from valves opening or shutting with a slow motion."

The question raised on this motion is, not whether Mr. Corliss is the inventor of these great improvements, but whether Mr. Frederick E. Sickles invented them; and he by no means lays claim to the invention of them all. Nevertheless, it is due to Mr. Corliss that he should have his true position here, as claiming in good faith to be the inventor of the engine which is thus brought in question. He submitted this engine to the Patent Office, and it was duly patented. That patent is strong presumptive, but not conclusive evidence of his title to the invention.

While we admit that others have devised improvements in regulating the power of the steam engine, by lifting the puppet-valves, and have combined the governor with other parts of the machinery to effect that regulation, it is equally certain that neither now nor in either of the two trials at law through which Mr. Corliss's patent has passed, has an engine organized and arranged like his, working in the same way, and producing the same results, or any description, model, or drawing of such an invention, been found, after the most thorough search which must be presumed to have been made in the machine shops, manufactories and libraries of this country and foreign countries. Thus much may suffice for the excellence of Mr. Corliss's engine, and his title as its inventor.

The Complainants allege that certain parts of it were invented and perfected by Frederick E. Sickles, and patented to him in the Letters granted May 20, 1842. Our first task in regard to this question is, to ascertain the true construction of that patent. The parties differ on the construction of each of its two claims.

The Complainants insist that the first claim is not merely for the precise mechanical arrangement of certain particular elements, which are described, but for a *new method* of effecting the cut off, by lifting and tripping the valves to cut off the steam, and so that it includes the principle of all arrangements made for the same purpose, and effecting the same ends in substantially the same way. On the contrary, the Defendants insist that it is a claim merely for the combination of particular elements, found only in engines using puppet valves, which elements are: 1st. The valve stem; 2d. The spring on the lifter; 3d. The adjustable sliding piece, with its wedges—all these working usefully, only when combined with the water reservoir, which is the subject of the second claim. We think that our construction is sustained by a fair and just collation of the passages in the patent, which bear on the question.

Thus the *recital:* "I have invented certain *improvements*

in the *manner* of *constructing* and *arranging* the *apparatus* for *lifting* and *tripping the valves* of steam engines, and by which the steam can be *more readily cut off at any desired part* of the stroke than by *the means heretofore adopted.*" This language manifestly implies the previous existence of means, viz.: apparatus of the same general character, producing the same effect, but less readily than the new device, and the application of the new improvement to that apparatus which so previously existed. In other words, it implies that steam is *already* cut off at any desired part of the stroke, by an *apparatus*, which lifts and trips the valves, to which existing apparatus the *improvement in the manner of arrangement* is to be applied.

So, the description of the improvement:

"In the accompanying drawing AA, fig. 1 represents *a valve box* containing the puppet valves, which are to be *lifted* and *closed*, the construction of this part being such as is *well known to engineers and machinists.* B is the valve stem, passing through a stuffing box, C, on the bonnet D of the valve box. The valve stem is to be raised by the lifter E, which is acted on the *usual way.*"

The usual way here referred to is evidently the usual way in which the valve stem is raised by the lifter, in the designed operation of tripping the valves to effect the cut off of the steam.

So also the *claim:*

"Having thus fully described the nature of my improvements in the *apparatus for lifting and tripping the valves of steam-engines and of thereby cutting off the steam at any required part of the stroke*, and also of my improved apparatus for regulating the closing of the valves, what I claim therein as *new*, and desire to *secure* by letters patent, is: first, the *manner* in which I have combined and arranged" [what evidently before existed] "the *valve stem*, the spring upon the lifter, the adjustable sliding piece, I, with its wedges and inclined planes, and their immediate appendages so as to co-operate with each other, and to effect the tripping of the valves and the cutting off of the steam substantially in the manner" [that is to say, the *particular* manner] "set forth."

It seems clear that the claim confines the patent to the *manner* in which the patentee has combined and arranged: first, the valve stem; second, the spring on the lifter; third, the adjustable sliding piece, with its wedges or inclined planes and their immediate appendages, to wit: the standard and setting screw so as to co-operate with each other and effect (what had been effected by some other combinations and not in the same manner,) the tripping of the valves and the cutting off of the steam.

But our construction of this part of the patent involves the further proposition, that this improved manner of making and constructing the apparatus mentioned in the first claim operates usefully, only, when connected with the water reservoir patented under the second claim. We proceed to prove this.

The inventor recites thus his second invention :

"And also an improved *water reservior* and *plunger*, which serve to prevent the slamming of the valves in closing, and consequently to preserve them in good working order for a great length of time."

Again, he closes his description of his second invention thus :

"By means of this apparatus, the valves may be made to shut so silently as scarcely to be heard, while the retardation is so perfectly graduated as not to be accompanied by any sensible loss of time as it takes place in the last moment of their descent only."

It is evident that this effect is not only subsidiary but *absolutely* essential to the main purpose of the first inventor, viz.: the lifting and tripping of the valves more readily than was done by previously known arrangements.

And then, in his claim of the second invention, the patentee says :

"I also claim the manner of regulating the closing of the valves, and of effectually preventing them from slamming, by means of a water reservoir, furnished with a piston or plunger, attached to the lower end of the valve stem, and operating within an adjustable cup or secondary reservoir, so as to effect the purpose intended upon the principle and substantially in the manner herein described."

It is thus clear that the apparatus covered by the first claim, and the apparatus covered by the second claim were intended to be combined, and that they must be combined, to effect the main purpose, viz.: the cut off by lifting and tripping the valves. For, if the valves, when tripped, (as described in the first claim,) should then fall to their seats, without being checked by the device covered by the second claim, the slamming of the valves would be a defect outweighing all that was gained by the new manner of tripping; and so, the whole apparatus would be practically useless.

Again, the newly invented water reservoir is entirely useless, except when it is applied to the lifting and tripping of *puppet-valves* in the apparatus before known to machinists which it was intended to improve by the invention of a new manner of arrangement.

This latter point, in our construction of the first clause, has been judicially sanctioned. His Honor, Judge Betts, in an action on the patent several years ago, instructed the jury to inquire, "Whether the plan of picking up the valve and catching it, as described in the patent, *in connection with the second reservoir*, was the invention of the patentee." So also His Honor, Mr. Justice Woodbury, in an action on this patent in the Rhode Island District, instructed the jury as follows:

"The plaintiff's patent is for a combination of particular elements, namely—the *valve stem*, the *spring* on the *lifter*, the *adjustable sliding piece, with its wedges*, arranged in the manner described, *working usefully only when combined with the water reservoir* described; and to find an infringement the jury must find in the defendant's *each one* of these elements, or a mechanical equivalent therefor, and they must be arranged and operated substantially in the same way as described in the plaintiff's specification."

The construction thus adopted by the learned Judge is sustained by the experts on this occasion. Truman Cook, one of the Complainant's witnesses says:

"These two parts cannot be used separately and produce a beneficial result—I mean an equally beneficial result. The tendency of the separation would be attended with uncertainty of action and increased liability to derangement of the parts."

William C. Hibbard, one of the Defendant's witnesses, says, (p. 3, quest. 6):

"Will the first part of the combination of machinery of the three elements referred to, work usefully except in connection with the water-reservoir?"

Answer—"Not as combined in the engine described; as the valves would otherwise strike so hard upon their seats as to destroy them in a short time."

So, also, William Raymond Lee, one of the Defendant's witnesses, (p. 41) answers the same question thus:

"It will not, as exhibited in model B. The arrangement involves a *drop* valve; and, without some resistance, fluid or otherwise, the valve would suffer material injury in being permitted to drop into its seat."

Thus we have sustained our construction of the first claim in Sickles's Patent.

The Complainants contend that the second claim (which need not be repeated here) covers a *method* of preventing the striking of the valves, by means of a reservoir, containing water, or any other fluid, compressible or incompressible, into which a plunger or piston, to which the valve stem is attached, enters, and, preventing the escape of the fluid is itself thereby arrested. On the contrary, the Defendants insist that the claim is for a mere technical combination of the piston or plunger attached to the valve stem with a reservoir of water, oil, or some other fluid, like them incompressible, applied and operating for the sole purpose of checking at a certain stage the descent of puppet valves in cutting off steam, and thereby preventing them from slamming on their seats.

We think that our construction is the one that is legitimately derived from the language of the patent. First, the recital :

"And also an improved *water-reservoir* and plunger, which serves to prevent the slamming of the valves in closing, and consequently to preserve them in good working order for a great length of time."

So, in the *description* of this invention :

"The reservoir J. is to contain *water*, *oil*, or some other fluid, say to *two thirds of its height*, *more or less*. Through the plunger K holes G G are represented as being made for the passage of *water* and H is a valve-like piece which slides up and down in the lower end of the stem B. This part of the apparatus, however, may be varied in its form in numerous ways *the intention being* to cause the *water* to offer a determined degree of obstruction to the descent of the plunger, and to admit of *this* being regulated. This I have sometimes done by making the plunger K a flat disc, with a sufficient space between it, and the cavity of the cup L for the passage of a portion of water sufficient to allow of the descent of the plunger, while it shall be so obstructed as to take off the force of the blow of the valve. I have, in fact, essayed the action of the plunger and of the parts within which it operates, in different forms, and in all with good effect, that which I have represented being one of the best. An opening furnished with a stopper may be made through the reservoir, as at I, to supply *water* when requisite."

So the claim, as has been already seen, limits the invention of the reservoir to a water reservoir, &c., with an "*adjustable* cup or secondary reservoir," operating upon the principle of regulating by the provision for the supply and escape of water, the resistance offered to the descending plunger or piston.

The term fluid, in its generic meaning, embraces compressible as well as incompressible fluids, not only water, oil, alcohol, and quicksilver, but also atmospheric air, the electric fluid, and (if there be such a thing), odyle or odic fluid. No one can read this patent without seeing that a cushion of air, with its appropriate devices, was no more in the minds of the inventor than an electric cylinder, a galvanic battery or a spiritual medium.

We close our remarks on the construction of the patent by observing that the broader pretensions made by the Complainants, cannot safely be upheld, because, as will appear in the sequel, methods for cutting off steam by lifting and tripping the valves, as well as for the employment of a water reservoir to check the descent of the valves and prevent their slamming, were well known in the arts, and were applied to the puppet-valve gearing in use when the patent was issued.

Assuming that our construction of the patent will be sustained, we proceed now to show that it is not infringed by the Corliss Engine.

1st. The puppet valve gear is essential to the use of Sickles's improvement. His patent is not infringed; because the puppet-valve gear, of which only it is an infringement by being combined with it, is not found in Corliss's engine. While the devices of Sickles's are expressly combined with the customary puppet-valve gear, and have no features of adaptation to engines using sliding or circular valve-gear, the former is absolutely excluded with all its parts, and the latter alone is found in the Corliss Engine.

2d. Analytical comparison of the two engines will show not only that Corliss has not adopted all of Sickles's parts, in their combinations, but even that scarcely one of those parts is found in the Corliss Engine; and so, on the contrary, that not one of the parts essential in the Corliss Engine is found performing the same function with the same result in the steam engine as organized by Sickles.

Certainly in Corliss's Engine there is, first, no "valve-box," (such as Sickles's patent requires,) "containing the puppet valves, which are to be lifted and closed," and the construction of which is well known to engineers and machinists." But there is, on the contrary, a steam chest, containing sliding or circular valves, which are not to be lifted at all at any time, and cannot be *lifted* from their seats, but can be removed from them only by a horizontal or sliding motion, and which are closed, not by falling on their seats, by pressure or gravity, but by a horizontal or circular motion.

Again, there is in Corliss's engine certainly no "valve stem passing through a stuffiing box on the bonnet of the valve box,' constructed in the manner known to engineers and machinists; that is to say, with one steam valve and one exhaust valve moved at one and the same time, and with a corresponding motion. But, on the contrary, each steam valve and its corresponding exhaust valve has its own separate and distinct lever which moves it by a separate motion and opens and closes it by a sudden movement, while the lever itself is never at rest; and which movements are not vertical and rectilinear movements, as in Sickles's engine; but, on the contrary, proceeding from different centres, act laterally and logitudinally, and not in a right line in any part of their operation. And thus the movements of these levers produce results not attained by Sicklcs's valve stem, to wit, the sudden opening and the equally sudden closing of the valves.

Of course, in Corliss's engine, "the valve stem is" *not* "to be raised by the lifter, which is acted on in the usual way;" for, in that engine, as there is no valve stem to be lifted, there is no *lifter* to raise it in any way.

Again, as there is in Corliss's engine, no lifter at all to be upheld, so, of course, there is no "spring," "attached to the shaft of the lifter," with "outer ends" "which embrace the sides of the valve stem." As there is, indeed, no valve stem at all, in Corliss's engine, so there is of course no "upper end of this stem," "flattened, or having projecting edges or feathers which, while the valve is being lifted, rest upon the upper edges of the spring where said feathers terminate."

Again, as there is in Corliss's engine no spring at all to open or shut, and no stem at all to be maintained or to be let fall, and no valves attached to any such stem, and to be closed by its motion, so there is no meaning at all in the words of Sickles's patent, "when the spring is opened the stem will no longer be sustained by it, and will consequently descend," if they are applied to any part of the mechanism of Corliss's engine.

Again, there is in Corliss's engine no "standard rising vertically" (or in any other position) "from the valve box or bonnet, so as that its upper flat end shall be nearly in contact with the out ends of the springs." And, as there is no such standard at all, so there is no "adjustable sliding piece" sustained by that standard, which may be shifted to any desired height, and then held in place by means of a set screw. Nor is there any set screw, nor any equivalent for it in the whole engine.

Again, as there is in Corliss's engine no adjustable sliding piece at all, so, of course, there are no "two projecting wedge-formed pieces or inclined planes, which serve to open the ends of the springs on the lifter, and thus to liberate the stem." And, as the sliding piece is absolutely wanting in Corliss's engine, so it is absurd to apply to that engine the directions contained in Sickles's patent for reversing the position of "the sliding piece, with its wedges or inclined planes," and so producing the same effect in a different form of operation.

Again, as there is in Corliss's engine no standard at all, so, that absent standard cannot be graduated at all.

Again, as the valve stem and the detaching spring and the lifter described by Sickles are all wanting in Corliss's engine, and as there is no equivalent for the lifter, so there is in that engine no "spring" "situated on the upper side of the lifter to which it is attached by one end, while its other end bears upon the stem," and operating to cause the stem to descend when the spring is opened.

So as there is not only not found in Corliss's engine the entire combination of the elements used by Sickles for effecting the cut off, but not even one of its parts, not one of the effects obtained by means of the apparatus devised by him is produced in the same way, or in the same degree, or to the same extent. The valves are not "tripped at any time during the ascent or descent of the lifter," (there being no lifter,) and the steam is not "cut off at any part of the stroke of the piston;" but, on the contrary, as we shall see in the sequel is cut off only during the half stroke, and by an apparatus entirely different.

So again, as there is no standard and no adjustment by the engineer, in Corliss's engine, the cut off is not and cannot be *adjusted* instantaneously, or otherwise, at any required point—that is to say, any point determined on and fixed by the engineer; but, as will be seen in the sequel, the regulation being purely automatic, the cut off has no such adjustability, but operates at points continually varying with every change in the heat of the furnace and the load of the machinery, which are the two ever-varying conditions by which the velocity of the engine is affected.

So of the second part of Sickles's patent—the water reservoir and its combination. There is in Corliss's engine no water reservoir. As there is no valve stem, so there is of course no "continuation of the valve stem" having affixed to it on its lower end a plunger or piston. Nor is there any plunger or piston at all.

As there is no plunger or piston to descend into a water reservoir, and no water reservoir at all, so the absent water reservoir has not " within it a cup, or secondary reservoir, which may be raised or lowered by means of a graduating screw, said screw being furnished with a nut, by which it is held in its place." Nor is there a graduating screw or a nut or an equivalent for either in the whole machine.

As there is no water reservoir, and no plunger or piston to descend into one, so the absent reservoir does not " contain water or oil or other fluid," "to two-thirds," or any other portion " of its height." Nor has the admitted plunger " holes for the passages of water " nor a " valve-like piece which slides up and down on the lower end " of the omitted stem.

Consequently, as this entire device, with all its apparatus, is wanting in Corliss's engine, no such operation takes place, and no such effect is produced there as those described in Sickles's patent. No valve-stem descends by means of any spring, or by force of gravitation, and so none is retarded in its descent near the point where the valves reach their seats. No slamming of the valves is prevented by the combined plunger and reservoir; and although none takes place yet it is because, from the very organization of the machine, none ever could take place.

On the contrary of all this, the valves in Corliss's engine, being sliding valves, are moved over their seats, not by a spring, or by their own gravitation, but by weights, with a motion not retarded by any water reservoir and plunger, or any other device, but accelerated and sudden; and this motion continues until after the valves are closed, and then the jar which would result from the sudden stoppage of the weight is prevented by air-cushions which receive the weights, and for which an India-rubber bed would be an exact equivalent.

Two things that are each equal to one and the same other thing are equal to each other. No one denies that Corliss's engine is described accurately in his patent. If that engine contains Sickles's elements in the same forms of combination, working in the same way, and working out the same results, we shall find them described in Corliss's patent.

Look at his *recital:*—

"In that class of steam engines in which the steam and exhaust ports of the cylindar are opened and closed by slide valves, whenever the valves close the ports, the steam presses them upon their seats with its whole force, and they cannot be moved without the expenditure of a considerable amount of power; but when the valves do not completely close their ports, the steam pressing upon both sides of them, does not tend to hold them upon their seats, and at these times the valves can be moved with but a small exertion of force. When a valve has closed its port, its office is performed; and hence the force exerted in any further movement of it while the port remains closed is wholly lost. Now it is customary in this class of engines to connect the valves rigidly, so that when one is moved the other is forced to move with it to the same extent; the closed valve is therefore moved with the opening one, and consequently the whole amount of force consumed in moving it while closed, is expended to no good purpose, and tends only to increase the wear and tear of the engine. To avoid this sacrifice of power, and at the same time to retain the advantages which result from the connection of the valves, is the object of the first part of my invention—which consists of moving each of the steam and exhaust valves of an engine independently by means of one crank wrist, of a series, which are all attached to a common disc, wrist plate, or other equivalent device which is secured to and moved with a rock shaft."

Here are described a class of engines and a valve gear to be improved, entirely different from those which Sickles describes, and to which he applies his improvement. The defects to be supplied and the advantages to be gained are entirely different from any of those which are the subject of any of Sickles's patent, and the manner indicated for supplying the one and governing the other are equally different. Sickles's object is to obtain a cut off at any part of the stroke, by lifting and tripping the valves of the puppet valve engine, and at the same time prevent their slamming on their seats. Corliss's is to save the power which is wasted in opening and closing the sliding valves by reason of their rigid connections. Sickles proposes to effect his purpose, by improving the rigidly connecting valve gear, which operates by the same and an equal motion on both valves at the same time. Corliss invents a new crank wrist with a separate lever, and a distinct motion to operate on each sliding valve. Mark, now, the operation. Each crank wrist is fixed on a different centre of motion, and imparts to the lever a motion that is neither vertical nor rectilinear, and which expends different measures of force at the same time on the two valves. Sickles's valve stem has but one rectilinear vertical motion, and imparts exactly the same measure of force to the two valves at the same time. In Corliss's engine—

" The several wrists which work the different valves are arranged upon the wrist plate in such positions with respect to the rods and levers or other devices which connect them with the valves, that they shall act like so many cranks, each of which vibrates near its dead point or point of slowest throw, and therefore imparts but little movement to the valve it actuates, when the latter is closed; while each moves with its fastest throw, and therefore communicates the greatest movement to its valve when the latter is open."

Contemplate, now, the effects thus obtained by Corliss's—

"Two great advantages result from this method of working the valves. In the first place, much of the power heretofore expended in moving the closed valve is saved: and secondly, the wire drawing of the steam is reduced, because the valves while opening and closing their ports are moved with increased speed."

Here are two new and useful results, neither of which is produced by Sickles, in his arrangement, or was ever conceived by him. Thus far, certainly, the inventions are entirely different.

Take, now, Corliss's second invention :

"The second part of my invention relates to the method of regulating the cut off of the steam in its passage into the engine, and consists in effecting this by means of the governor, which operates cams, so that when the velocity of the engine is too great, these cams shall be moved by the action of the regulator to such positions that catches on the valve rods may the sooner come in contact with them to liberate the valves, and admit of their being closed by the force of weight or springs, and thus cut off the steam in proportion to the velocity of the engine; this being done sooner when the velocity of the engine is to be reduced and later when it is to be increased."

Here follows a minute description, by which it is seen that while in Sickles's engine the engineer makes a fixed adjustment at pleasure without the aid of the governor, or of any combination of it, and leaves the governor to regulate by wire-drawing the steam at the throttle-valve, Corliss's has an automatic variable adjustment, by withdrawing the governor from the working of the throttle-valve, and, indeed, dispensing with the throttle valve altogether, and by combining the governor with the catches that liberate the several levers which work the valves—a new combination, not found in Sickles's engine, working in a way entirely different from his, and producing results that he has not attained, nor even conceived. I read the description :

"In the Steam Engine represented in the accompanying drawings, the steam and exhaust valves *l*, *l*, and *m*, *m*, are situated in steam chests *n*, *o*, at each extremity of the steam cylinder; the chest *n*, at the top is formed in the cylinder head, while the other *o*, is let into a recess in the bed-plate. Each exhaust valve *m*, is attached to one extremity of a valve rod *p*, which is fitted at its opposite extremity with a sliding head *q*, that is linked by a connecting rod *r*, to one arm *s*, of a bell crank *t*. The other arm *v* of the bell crank is connected by a rod with a wrist pin *w* on the wrist plate *x*. The latter is secured to a rock shaft to which the requisite vibratory motion is imparted by an eccentric *a* through the intervention of an eccentric rod and an arm *y* secured to the rock shaft.

"The wrists *w* of the two valves are in this example a quarter of a circle distant from each other, and the two connecting rods extend in opposite directions from the rock shaft—hence, when one wrist is at its point of greatest throw, the other is at its dead point; and when one is imparting to its rod and the valve connected therewith, the greatest movement, the other is imparting to its valve the least. Each valve is therefore moved alternately fast and slow; and the fast movement of one is effected during the slow movement of the other; nearly the whole movement or throw of each valve being effected while the port is either partially or wholly open, at which time the least power is required to move it; while as the small remnant of the throw when the port is closed is effected during the slow movement, but little power is then required, as the distance to which the valve is moved is now very short. The steam valves *l*, *l*, are worked in a manner similar to that of the exhaust valves, with the exception of an arrangement by means of which they are made to close and cut off the supply of steam at any required portion of the stroke. The valve rods *b* of these valves are double, and instead of being permanently linked to their appropriate bell-cranks, are each connected by a detachable link *h'* with a rack *g'*, whose teeth engage with those of a tooth sector *f'* on the bell-crank. This detachable link *h'* is hinged at one extremity to the cross head *c'* which unites the two members of the double valve rod; it has a shoulder *i'* at its opposite extremity which engages in a corresponding socket on the rack, and is kept in place by a spring *i''*. This link *h'* is also fitted near the rack with a projection *j'* which is struck at the proper moment to detach the link from the rack by a revolving helical cam *k'*. The helical cams which detach the links *h' h'* of the two steam valves *l*, *l*, are both secured to an upright shaft *l'*, which is caused to revolve by the movement of the crank shaft, and is arranged at the same time to move freely up or down in its boxes. This shaft is connected at *r'* with the governor (which in the present instance is of the centrifugal variety, that being the kind I have used, deeming it the best,) so that it shall move up or down as the balls of the governor rise and fall. When the governor balls *s' s'* are at their lowest position, the shaft *l'* is depressed so far that the helical cams *k' k'* are below the range of the valve link, *h' h'*, and consequently cannot detach them; hence in this position of the governor balls, the steam valves, being connected with their wrist pins throughout the whole length of the stroke, are opened and closed in the same manner as the exhaust valves. As, however, the velocity of the engine is increased, and the governor balls rise under the increased centrifugal force, the upright shaft *l'* is correspondingly raised, and the cams being now revolved within the range of the valve links, strike the projections *j*, and detach the links *h'* from their racks. The helical cams extend round the shaft in the same direction as the latter is turned, consequently the higher the shaft *l'* and its cams are raised the sooner will the cams strike the projections and detach the links. As soon as the links are detached, the valves being entirely disconnected from the mechanism by means of which they are opened, are consequently free to close. A convenient mode of arranging the several parts when the valves move parallel with the axis of the cylinder, is represented at Figs. 8 and 9, in which the letters indicate the parts corresponding with those indicated by them in the arrangement before described. I wish it to be distinctly understood, that in the mode of regulating the cut off by the governor, I do not limit myself to the use of the particular kind of cams described or represented; as the form, position, and operation of these, may be greatly varied without changing the principle of this part of my invention; as for instance, stops or cams connected with the slide of the governor by levers, may be made to slide in the direction of the plane of motion of the valve rods, to vary the periods of liberating the catches of the valve rods; or, wedge formed stops or cams may be substituted for the helical cams and attached to the cam rod, which, in that case, must not turn."

Proceeding through the description, the reader finds the devices of weights, a socket and an air cushion, which present a faint resemblance in form to the plunger and water reservoir of Sickles' engine. But it appears, on examination, that the devices are entirely different, the mode of operation equally different, and the effect attained as different as the checking of a motion at a point which involves a loss of power to the engine, is from the giving the motion full way until all the power has been effectively applied and then providing for its sinking to rest. This part of the description is as follows:

"As the steam valves in the steam engine represented move horizontally they do not tend to close by their own weight, and are consequently closed by means of weights o', o', which act through the intervention of bent levers, or bell cranks, m' upon cross-blocks n'', secured to the respective valve rods. In order to prevent the jar which would result *from the sudden stoppage of the motion of the weight*, each cross-block, n', has a cylindrical socket formed in the face near the steam cylinder, and a piston, p', is secured to the engine frame which enters the cylindrical socket and compresses the air within it to form an elastic cushion to prevent the jar. As the racks g', are moved back by the action of the wrist pins and the bell cranks, the shoulder of the link re-engages with the socket on the rack, so that the valve being now re-connected with valve mechanism, is opened by the wrist pin at the proper moment to admit steam into the cylinder."

This contrast of the two engines cannot be more effectually completed than by presenting in parallel columns the claims of the respective inventors:

CORLISS.

What I claim as my invention and desire to secure by letters patent is, first, the method substantially as described of operating the slide valves of steam-engines by connecting the valves that govern the ports at opposite ends of the cylinder with separate arms of the rock shaft or the mechanical equivalents thereof; so that from the motion thereof the valve that keeps its port or ports closed shall move over a less space while its port or ports is closed than the one that is opening or closing its ports, and vice versa; while at the same time, the two arms by which they are operated, have the same range of motion as described; whereby I am enabled to save much of the power heretofore required to work the slide valves of steam engines, and by which also I am enabled to give a greater range of motions to the valves at the periods of opening and closing the ports to facilitate the induction and eduction of the steam as specified.

And lastly, I claim the method of regulating the motion of steam engines by means of the Regulator, by combining the said Regulator with the catches that liberate the steam valves, by means of moveable cams or stops, substantially as described. GEORGE H. CORLISS.

SICKLES.

Having thus fully described the nature of my improvements in the apparatus for lifting and tripping the valves of steam engines, and of thereby cutting off the steam at any required part of the stroke and also of my improved apparatus for regulating the closing of the valves; what I claim therein as new and desire to secure by letters patent is, first, the manner in which I have combined and arranged the valve stem B, the spring F on the lifter, the adjustable sliding piece I, with its wedges or inclined planes and their immediate appendages so as co-operate with each other and to effect the tripping of the valves and the cutting off of the steam substantially in the manner set forth. I also claim the manner of regulating the closing of the valves and of effectually preventing them from slamming, by means of a water reservoir furnished with a piston or plunger attached to the lower end of the valve stem, and operating within an adjustable cup or secondary reservoir, so as to effect the purpose intended upon the principle and substantially in the manner herein described and made known.

FREDERICK ELSWORTH SICKLES.

If Corliss's engine contains Sickles's improvement, and if each of the two specifications be sufficient, then it follows that a mechanic ordinarily skillful in the art of constructing steam engines could, without any exercise of invention, produce Sickles's devices, in pursuance of directions contained in Corliss's patent, and Corliss's engine by following the directions contained in the patent of Sickles. Where, among all the complainant's experts, men of extraordinary skill, has been found the one who is willing to say that, following the directions of the one patent, he could, without invoking invention, produce the engine described in the other? There is not one. And well must it be so, for it has been seen that, from the main eccentric to and including the valves, and including everything which concerns the working of the valves, everything in Corliss's engine is new and different from that of Sickels, and from the steam engine as modified by his improvements.

The complainants, however, insist that in Corliss's engine they find, in the wedge like position of the levers which move the valves against the regulating stop, the equivalent of the adjustable slide, with its wedge-like faces or inclined planes of Sickles. On the contrary, it is seen that those levers, although they are in an inclined position, do not at all act as wedges. When acting so as to liberate the hooks, they move in exactly the opposite direction from that in which the wedge moves when in action, and they liberate the hooks as levers by coming to a bearing on the ends of the adjusting stop as a fulcrum.

Let us now briefly bring into more distinct view the radical differences between the inventions already described. Sickles's cut off is one adjustable by the skill or manual interpositon of the engineer, and invariable except by readjustment. Corliss's is variable and purely automatic. Sickles's cut off is imperfect because dependent on the engineer, who can only approximate to accuracy. Corliss's cut off is practically perfect. Sickles's cut off uses a governor to regulate at the throttle-valve. Corliss's, on the other hand, combines the governor with the steam-valves. Corliss's brings the valves, when detached, quickly and suddenly over their seats. Sickles's retards the movement just before they reach their seats. Sickle's loses part of the expansive force of the steam at the throttle valve and further parts at the steam valve. Corliss's, by dispensing with the throttle valve, and opening and closing the steam valves by quick and sudden motions, saves and works the whole expansives force of the steam. Sickles's cut off may be adjusted so as to take effect at any part of the stroke of the piston, which may be a desirable improvement in marine engines, in cases of stormy weather and rough seas, while Corliss's can only be made to take effect during the half-stroke, which is all that is desired in stationary engines. Finally, as might be expected in engines organized with a view to results so widely different, the mechanism of the one is entirely different from that of the other.

So it results that, regarding the parties with equal favor, there is no occasion for complaint by one against the other; no ground of conflict between them. They are competitors for a prize that both may win.

Here, if the responsibility of counsel would allow, we would rest. But this responsibility requires us to show, further, that Sickles's alleged improvements were not new.

In the first place let the construction of the patent be as insisted by the complainants, viz: that the patent is for methods of cutting off steam by tripping the valves or detaching them from the mechanism which moves them and of checking the descent of the valves, so as to prevent their slamming. Then the patentee was not the inventor of these methods, for such methods existed and were known before.

Here is our evidence as to the first claim—Tredgold, Art. 478.

"Modes of opening and closing valves, cocks and slides. The motion may be given either from the reciprocating or from the rotary parts of an engine. In engines which have no rotary parts, motion is communicated to the valves by a rod or beam, called a plug tree, attached to the engine beam near to the end moved by the piston rod. This plug tree is provided with certain adjustable projections called tappets, which strike the levers or handles of the valves, and thus open or shut them at the proper intervals as the beam ascends or descends. These handles turn on axis and act as levers to move the valves, slides or cocks * * * * When valves are employed they are generally opened by weights."

(See plate 9, fig 3, where is exhibited the device described in the text.) The author proceeds:

"The descent of the weight which opens a valve is regulated by an ingenious method. It either descends into or forces a piston into a vessel of water (see C. Fig. 3, Plate 9,) while the aperture by which the water escapes from under it may be increased or diminished at pleasure. The weight, therefore, acts with its full force to open the valve, but as soon as it begins to move, it is retarded by the water until it is finally stopped. During the ascent a valve opens inwardly at the bottom of the vessel, and therefore the engine had not more than the weight to raise again."

Certainly this was a method of checking the descent of valves and of preventing their slamming by the device of a water reservoir and piston.

Again, here is a valve gear used by Watt. Stuart's Anecdotes of the Steam Engine, p. 366, Vol. 2:

"The different construction of Watt's valves required a new arrangement of the hand-gear to move them. In the atmospheric engines the weight or tumbling bob at the end of levers, called the Y piece, was placed there, so that its fall should open the sliding valve or the injection cock with a jerk. It exerted no influence whatever in keeping the valves in their places, at least it was not introduced for that purpose. Watt also sometimes attached a weight to the spanners which opened or shut his valves. But it was necessary in some arrangements of these parts that they should be gradually shut to prevent the injury that would be occasioned by a sudden jerking of the valves into their seats. This weight is therefore *sometimes formed like a piston which rose and fell in a cylinder with water.* The weights or plungers are made of cast iron, and cylindrical, each fitted into a hollow cylinder in the condensing cistern, and covered with water. The plunger is made somewhat less than the barrel to allow a small space, by which, when it descends, the water may rise between it and the barrel. The lower end of each barrel is closed, except a small hole, which is covered by a leather valve opening inward. When the plunger is drawn up, the water from the cistern flows through this valve into the barrel; but when the plunger descends the valve closes, and the water which is displaced rises between the plunger and the barrel, and the resistance which is thus occasioned to the descent of the weight prevents the concussion which would be produced by its uninterrupted fall."

Then follow references to a plate giving a full view of the machine. Thus we find here the machine actually used by Watt, so far as any principle is involved, containing the two identical methods of effecting the indentical purposes contemplated by the patent of Sickles, for which, according to the broader construction of the complainants, the patent issued.

At most, the only difference between Watt's and Sickles's respective methods for checking the descent of the valves by a plunger and water reservoir is, that Watt's began the retarding motion at a point in the stroke somewhat earlier than Sickles's. But the combinations and effect were the same, and both were applied to substantially the same purpose, viz.: to prevent the slamming of the valves in closing. In fact, it may be assumed that ever since Watt invented the weight to detach the puppet valves from the mechanism that moved them, substantially the same method has been used to prevent them slamming on their seats.

Again, still considering the patented things as *methods*, both were known to and used by Phineas Bennet in 1838. Your Honor will please notice that Sickles, before obtaining his patent, on filing his caveat, knew of Bennet's device, whatever it was. (Deposition of Phineas Bennet, pp. 54, 55, question 23.) But, moreover, the devices known and used are the same as those of Sickles. (Bennet's Deposition, p. 52, questions 3, 4, 5, 6, 8, 9, 10, 11. See, also, the Deposition of Livingston S. Bartholomew, p. 62, question 3 to 8, inclusive. See, also, the Deposition of Ezekiel Williams, p. 68.)

These were not experiments, but a perfectly successful and established result, as all the witnesses show; Bennet's differing only in mechanical arrangement, but the methods, as we see, were the same.

And now, assuming, according to the Defendant's construction, that the patent covers only the technical mechanical combinations, and not the method or principle, *even then the inventions are not new.* Exactly the same were known and used on the South America, built in 1840 and 1841 by Hogg & Cunningham. (See the Depositions of James Hogg, beginning at p. 70; of George Hawes, p. 88; of Cornelius H. Delamater, p. 94; of Joseph Belknap, p. 105; of Joseph Birkbeck and of Robert Robertson, p. 132.)

The only difference between Hogg's cut off and Sickles's was that Hogg used a standard of different form and had not the same adjustable sliding piece as Sickles's has, but effected the disengagement of the valve stem, by a fork that rested on the lifter, and worked into notches on the valve stem, and was drawn out of the notches by a stationary cam as the lifter ascended and was pressed back into notches by a spring on the lifter, as the lifter descended and the mechanism for reversing lacked, so as to cut off on the last part of the stroke in which respect it is like that of Corliss's. But I need not argue this, for the reason that the complainants admit that Hogg's device is identical with that of Sickles's. The water reservior was the same, and had a secondary reservior operating in the same way. While Hogg shows that he made the invention in 1838, all the witnesses named show that it was completed and successfully applied in April, 1841. Sickles's patent is dated May 20, 1842, and the letter by which he claims that he suggested the improvement to Hogg is dated in May, 1841. It is apparent, moreover, that at the date of that letter, Sickles had not conceived the secondary reservoir described in his patent, for this letter does not mention it. Sickels's earliest date of his invention, as proved by William B. Sickles on the Rodman trial was in 1838. Benjamin R. Brown says, in July or August, 1839. Frederick Sickles says in cold weather, 1839.

Moreover, it is apparent that Sickles obtained his patent through a gross mistake of facts or by perjury. For the patent was refused at first on the ground that the inventions were used on the South America (before she broke down) in April, 1841; and then his partner Truman Cook sent a supplemental affidavit asserting that they were not so used on the South America before she broke down. Cook testifies that the patent was issued *on that* affidavit, (p. 2, vol. 1) whereas it is now shown by proof that the devices were in successful operation on the South America before that accident; and this, too, is established by the following witnesses; Peter Hogg and Cornelius H. Delamater; George Birkbeck, Jr., steam-engine builder; George Hawes, formerly engineer of People's Line of Steamboats; Robert Robertson, Chief Engineer of U. S. Mail Steamship Baltic. And, by comparing Robertson's testimony with Sickles's caveat, it is manifest that Sickles must have obtained his ideas of the invention from the South America.

Surely I need not dwell on the point that the Complainants have not established their title by a verdict, or by undisturbed possession. Their verdict against Rodman was not on the present issue. If it was it is balanced by the effect of the patent of Corliss afterward granted. And in a trial since, in the Rhode Island District, the Complainants failed to obtain a verdict altogether, owing to defect of pleading or to deficiency of proof, or errors in the conduct of the case, they obtained a verdict against Rodman, in 1843, sustaining the Complainant's title; yet Hogg & Delamater have been making and selling the devices covered by Sickles's patent; and that, too, unmolested.

In conclusion, even if we have not conclusively established our defense, we have, at least, established it sufficiently to entitle us to a trial at law on a feigned issue. And to that trial we challenge the Complainants, who alone can give it, unless the Court shall see fit here to order it, for the purpose of ascertaining the facts necessary to guide its conscience on the final merits of the controversy.

It may well be doubted whether in the entire course of the administration of the Patent Law, a bolder experiment was ever made to pervert it to the purposes of injustice and oppression. The Complainants, without the full title which the law requires, without that undisturbed possession which the law requires, without having established the title to the invention before a Jury as the law requires, and without having established the fact of infringement by the verdict of a Jury, as the law requires—prosecute a party on a patent, which, if construed to cover the method of cut off by tripping the steam-valves, as they claim it shall be, is as old as Watt; and, if construed as a patent for a *mere device*, was invented, matured, and perfected by two citizens of this State, eminent in their business, a year before the patentee claims to have made his invention and whose title to the discovery, although not presented by themselves at the Patent Office, was ascertained there, and made a ground for rejecting the Patent, and was only set aside and overruled there, on an exparte deposition of an interested party, now proved to have been erroneous and false. To indulge apprehension in such a case would be to impeach by our doubts the wisdom and the justice of the tribunals of our country.

www.ingramcontent.com/pod-product-compliance
Lightning Source LLC
LaVergne TN
LVHW011128110826
845150LV00008B/2276

* 9 7 8 1 4 1 8 1 9 5 0 8 3 *